Marco Antonio González Morales

Linear Algebra for Teaching Activities

Marco Antonio González Morales

Linear Algebra for Teaching Activities

Vol. II

SciénciaScripts

Imprint
Any brand names and product names mentioned in this book are subject to trademark, brand or patent protection and are trademarks or registered trademarks of their respective holders. The use of brand names, product names, common names, trade names, product descriptions etc. even without a particular marking in this work is in no way to be construed to mean that such names may be regarded as unrestricted in respect of trademark and brand protection legislation and could thus be used by anyone.

Cover image: www.ingimage.com

This book is a translation from the original published under ISBN 978-613-9-40621-0.

Publisher:
Sciencia Scripts
is a trademark of
Dodo Books Indian Ocean Ltd. and OmniScriptum S.R.L publishing group

120 High Road, East Finchley, London, N2 9ED, United Kingdom
Str. Armeneasca 28/1, office 1, Chisinau MD-2012, Republic of Moldova, Europe
Printed at: see last page
ISBN: 978-620-7-97685-0

LINEAR ALGEBRA FOR TEACHING

VOL. II

Marco Antonio González Morales
marco.gonzalez@academicos.udg.mx

INTRODUCTION

The main objective of this material is to have a friendly and digestible narrative for all those colleagues who begin in the wonderful and enriching teaching career, teaching topics related to Linear Algebra for the area of engineering and that also serves the student to understand in a more friendly way in a less dense language in technical terms, the topics, concepts and methods of linear algebra.

Linear Algebra is an essential part of the mathematical training of every scientist, administrator and engineer, since its applications are numerous in the areas of physics, chemistry, biomedical engineering, computer graphics, image processing, process optimization, among many others.

In the content of this material you will find the following topics:

1. Matrices
2. Determinants

This is why reference is made to great authors of books focused on linear algebra, such as Baldor (Baldor, 2019) with authorships on Algebra, and other expert authors on the subject (David C. Lay, Ron Larson, Grossman, Estrada, Guzmán, among others) is why this project entitled ***Linear Algebra for the Teaching Activity*** was born. ***Vol. II***, which contemplates topics related to Algebra and the branch of Linear Algebra, in this second volume topics on ***Matrices*** and ***Determinants*** are addressed; with the demonstration of exercises related to the methods and an annex of exercises proposed for practice.

CONTENTS

WHAT ARE MATRICES AND DETERMINANTS IN LINEAR ALGEBRA?

In **Linear Algebra**, a **matrix** is a fundamental structure used to represent and manipulate numerical data in an organized manner. A matrix is essentially a two-dimensional set of numbers arranged in rows and columns. Each number in a matrix is called an "**element**" and is identified by its position in the corresponding row and column of the mode:

$$
A = \begin{pmatrix}
a_{11} & a_{12} & a_{13} & \cdots & a_{1n} \\
a_{21} & a_{22} & a_{23} & \cdots & a_{2n} \\
\vdots & \vdots & \vdots & \ddots & \vdots \\
a_{m1} & a_{m2} & a_{m3} & \cdots & a_{mn}
\end{pmatrix}
\left.\begin{matrix} \leftarrow \\ \leftarrow \\ \leftarrow \\ \leftarrow \end{matrix}\right\} \text{Filas de la matriz A}
$$

$$\underbrace{\phantom{a_{11} \quad a_{12} \quad a_{13} \quad \cdots \quad a_{1n}}}_{\text{Columnas de la matriz A}}$$

Matrices are generally represented using capital letters, such as **A**, **B**, **C**, etc. and are expressed as $A = (a_{ij})$. Each element of the matrix carries two subscripts. The first one, "i", indicates the **row** in which the element is located, and the second one, "j", the **column**. Thus the element a_{23} is in **row 2** and **column 3**.

If a matrix has **m** rows and **n** columns, it is said to be of **size mxn** or to have **mxn dimensions**. For example, a 2 x 3 matrix has 2 rows and 3 columns. (Larson, 2013)

Here is an example of a 2 x 3 matrix:

$$
A = \begin{pmatrix}
1 & 2 & 3 \\
4 & 5 & 6
\end{pmatrix}_{2 \times 3}
$$

Matrices are used in the context of science as elements that serve to classify numerical values according to two criteria or variables.

Some key concepts related to matrices in linear algebra include:

Elements: These are the individual values within the matrix, such as the numbers 1, 2, 3, 3, 4, 5 and 6 in the matrix above.

Rows and columns: Rows are arranged horizontally, while columns are arranged vertically. In the example above, the first row is [1, 2, 3], and the second row is [4, 5, 6].

Size or dimension: Refers to the number of rows and columns in a matrix. In the example above, the matrix has a size of 2 x 3.

Square Matrix: When a matrix has the same number of rows as columns, it is called a square matrix. For example, a 3 x 3 matrix is square.

Identity matrix: It is a square matrix in which all the elements are zeros, except those of the main diagonal (from top left to bottom right), which are all equal to 1.

Operations with matrices: Several operations can be performed with matrices, such as addition, subtraction, multiplication, transposition, inversion, among others.

Matrices are used in a wide variety of applications in mathematics, physics, computer science, statistics, engineering and many other disciplines. *They are a powerful tool for solving systems of linear equations, representing linear transformations, and performing calculations in linear algebra and numerical analysis*. (Grossman, 2019)

In contrast, the *determinant* is a numerical function associated with a square matrix. In linear algebra, the determinant of a matrix is a measure used in various applications, such as solving systems of linear equations, matrix inversion and determining the linear independence of a set of vectors. (Guzman, 2011)

The *determinant of a square matrix A* is typically denoted as *det(A)* or |A|, and is calculated in different ways depending on the size of the matrix. For a 2x2 matrix:

If A is a 2x2 matrix:

$$A = \begin{pmatrix} a & b \\ c & d \end{pmatrix}_{2x2}$$

The determinant of **A** is calculated as:

$$det(A) = (a \cdot d) - (c \cdot b)$$

For a 3x3 matrix:

$$A = \begin{pmatrix} a & b & c \\ d & e & f \\ g & h & i \end{pmatrix}_{3\times3}$$

The determinant of **A** is calculated by the "*Sarrus Rule*" as follows:

$$det(A) = a(e{\cdot}i - f{\cdot}h) - b(d{\cdot}i - f{\cdot}g) + c(d{\cdot}h - e{\cdot}g)$$

For larger matrices, such as 4x4 or larger, more advanced methods are used, such as *cofactor expansion* or *Laplace's Rule*.

The determinant of a matrix has some important properties:

- *The determinant of a matrix is zero if and only if the matrix is singular, which means that it has no inverse.*
- *The determinant of a matrix is the product of the determinants of its elementary factors, that is, if you can express a matrix as a product of smaller matrices, you can calculate its determinant by decomposing the calculation.*
- *The determinant of the identity matrix (a square matrix with ones on the main diagonal and zeros everywhere) is equal to 1.*
- *The determinant of a matrix changes when its rows or columns are interchanged.*
- *The determinant of a scalar matrix (a matrix with a single number everywhere) is equal to the number raised to the power of the number of rows (or columns) of the matrix.*

Determinants are useful in a variety of contexts, such as solving systems of linear equations, matrix inversion and determining linear independence of vectors. They are also used in determinant theory in advanced mathematics and in fields such as geometry and physics. (Lay, 2007)

CHARACTERISTICS AND TYPES OF MATRICES

A *matrix* is a set of elements arranged in *Rows* (rows) and *Columns*.

Examples:

$$A = \begin{pmatrix} 1 & 3 \\ 6 & 5 \\ 5 & 2 \end{pmatrix}$$ i.e. it is a 3x2 **ORDER** matrix.

$$B = \begin{pmatrix} 2 & 5 & 7 \\ 3 & 4 & 2 \end{pmatrix}$$ i.e. it is a Matrix of **DIMENSION** 2x3

It is important to mention that the word **ORDER** is synonymous with **DIMENSION**.

Each **ELEMENT** of the matrix is identified with the lower case letter corresponding to the matrix, i.e. the first element of **Matrix A** would be e_{11} , whose numbering$_{11}$ emphasizes the location of where it is, **i.e.** the element$_{11}$ corresponds to the first Row and first Column.

$$A = \begin{pmatrix} \textcircled{1} & 3 \\ 7 & \textcircled{2} \\ 3 & 4 \end{pmatrix}_{3x2} \quad a_{11}, \, a_{22} \qquad B = \begin{pmatrix} \textcircled{1} & 3 & 2 \\ 3 & 4 & \textcircled{5} \end{pmatrix}_{2x3} \quad b_{11}, \, b_{23}$$

A_{ij} $\qquad\qquad b_{ij}$

i = Filas
j = Columnas

Exercises:

$$A = \begin{pmatrix} 1 & 3 & 4 & 2 \\ 7 & 2 & -7 & 3 \\ 3 & 4 & 1 & 0 \end{pmatrix}$$

ORDEN = 3 x 4
a_{21} = 7
a_{34} = 0
a_{43} = No hay

$$B = \begin{pmatrix} 1 & 3 \\ 7 & 2 \\ 3 & 4 \\ 4 & 6 \end{pmatrix}$$

ORDEN = 4 x 2
b_{11} = 1
b_{21} = 7
b_{22} = 2
b_{31} = 3
b_{42} = 6

SQUARE MATRIX

A *square matrix* is one whose **Rows** and **Columns** *are equal*, i.e. *2x2, 3x3, 4x4, ... , nxn.* And the *concepts for its interpretation* are:

The **MAIN DIAGONAL** is the one formed by the elements at_{11} , at_{22} , at_{33} , ... at_{mn} .

$$A = \begin{pmatrix} -5 & -2 \\ 4 & 0 \end{pmatrix}_{2x2} \qquad B = \begin{pmatrix} -2 & 6 & 4 \\ 8 & 12 & -9 \\ 15 & -6 & -4 \end{pmatrix}_{3x3}$$

TRACE: is the sum of the elements of the **Main Diagonal**, respecting the signs of the elements, that is to say:

$$A = \begin{pmatrix} -5 & -2 \\ 4 & 0 \end{pmatrix}_{2x2} \qquad B = \begin{pmatrix} -2 & 6 & 4 \\ 8 & 12 & -9 \\ 15 & -6 & -4 \end{pmatrix}_{3x3}$$

$$A = (-5) + 0 = -5 \qquad\qquad B = -2 + 12 + (-4) = 6$$

UPPER TRINGULAR: is the matrix in which the elements <u>below</u> the main diagonal are all zeros.

$$A = \begin{pmatrix} -2 & 6 & 4 \\ 0 & 12 & -9 \\ 0 & 0 & -4 \end{pmatrix}_{3x3}$$

LOWER TRINGULAR: is the matrix where the elements <u>above</u> the main diagonal are all zeros.

$$A = \begin{pmatrix} -2 & 0 & 0 \\ 8 & 12 & 0 \\ 15 & -6 & -4 \end{pmatrix}_{3x3}$$

DIAGONAL MATRIX: is the matrix where the elements that ARE NOT in the main diagonal are zeros.

$$A = \begin{pmatrix} 5 & 0 \\ 0 & 4 \end{pmatrix}_{2x2} \qquad\qquad B = \begin{pmatrix} 3 & 0 & 0 \\ 0 & 8 & 0 \\ 0 & 0 & 4 \end{pmatrix}_{3x3}$$

- *Another characteristic of the diagonal matrix is that they are upper triangular and lower triangular at the same time.*

IDENTITY MATRIX: is a diagonal matrix in which **all the elements** of the **main diagonal are ones**.

$$M = \begin{pmatrix} 1 & 0 \\ 0 & 1 \end{pmatrix}_{2x2} \qquad\qquad N = \begin{pmatrix} 1 & 0 & 0 \\ 0 & 1 & 0 \\ 0 & 0 & 1 \end{pmatrix}_{3x3}$$

SCALAR MATRIX: is a diagonal matrix in which **all the elements** of the **main diagonal are equal**.

$$A = \begin{pmatrix} 5 & 0 & 0 \\ 0 & 5 & 0 \\ 0 & 0 & 5 \end{pmatrix}_{3x3} \qquad\qquad B = \begin{pmatrix} -4 & 0 \\ 0 & -4 \end{pmatrix}_{2x2}$$

Examples.

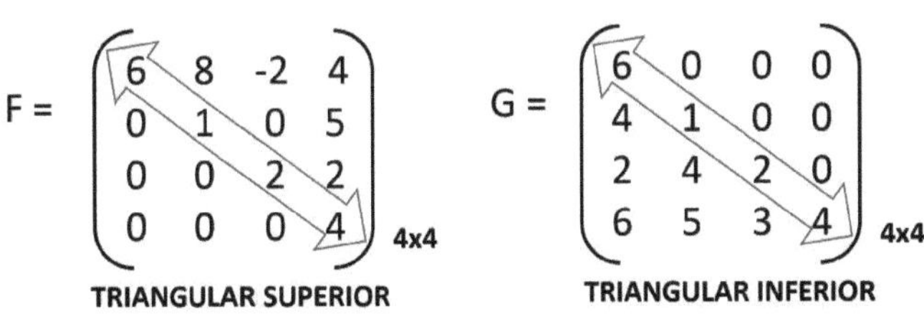

D =
3 0
0 3
2x2
ESCALAR

E =
1 7 3
5 4 2
8 2 1
3x3
MATRIZ CUADRADA

F =
6 8 -2 4
0 1 0 5
0 0 2 2
0 0 0 4
4x4
TRIANGULAR SUPERIOR

G =
6 0 0 0
4 1 0 0
2 4 2 0
6 5 3 4
4x4
TRIANGULAR INFERIOR

There are also different *types of matrices*, which can be identified as follows:

ROW MATRIX: it is also known as **ROW VECTOR**, which is formed by a single row.

$$A = \begin{pmatrix} a_{11} & a_{12} & a_{13} & \dots & 1_{an} \end{pmatrix}_{1xn}$$

Ejemplos...

$$\begin{pmatrix} 5 & 3 & 2 \end{pmatrix}_{1x3} \qquad \begin{pmatrix} -7 & 12 & 0 & 3 \end{pmatrix}_{1x4}$$

COLUMN MATRIX: it is also known as **COLUMN VECTOR**, which is formed by a single column.

Ejemplos...

$$A = \begin{pmatrix} a_{11} \\ a_{21} \\ a_{31} \\ \vdots \\ a_{m1} \end{pmatrix}_{mx4} \qquad \begin{pmatrix} 5 \\ 0 \\ -2 \end{pmatrix}_{3x1} \qquad \begin{pmatrix} -6 \\ 8 \\ 12 \\ -3 \end{pmatrix}_{4x1}$$

NULL MATRIX: is a matrix that all its elements are null (zeros).

$$M = \begin{pmatrix} 0 & 0 & 0 & 0 \\ 0 & 0 & 0 & 0 \\ 0 & 0 & 0 & 0 \end{pmatrix}_{3x4}$$

SQUARE MATRIX: it is made up of the same number of rows as columns.

$$A = \begin{pmatrix} -5 & 0 \\ 7 & 9 \end{pmatrix}_{2x2} \qquad B = \begin{pmatrix} 4 & 8 & -5 \\ 0 & 0 & 2 \\ 13 & -4 & -10 \end{pmatrix}_{3x3}$$

MAJOR DIAGONAL (DOMINANT): it is the one formed by the elements at_{11} , at_{22} , at_{33} , ... at_{mn} .

$$A = \begin{pmatrix} -5 & 0 \\ 7 & 9 \end{pmatrix}_{2\times 2} \qquad\qquad B = \begin{pmatrix} 4 & 8 & -5 \\ 0 & 0 & 2 \\ 13 & -4 & -10 \end{pmatrix}_{3\times 3}$$

MINOR DIAGONAL: it is the one formed by the elements a_{ij} , where $i + j = n + 1$.

$$A = \begin{pmatrix} -5 & 0 \\ 7 & 9 \end{pmatrix}_{2\times 2} \qquad\qquad B = \begin{pmatrix} 4 & 8 & -5 \\ 0 & 0 & 2 \\ 13 & -4 & -10 \end{pmatrix}_{3\times 3}$$

Examples.

$$A = \begin{pmatrix} 0 & 0 \\ 0 & 0 \end{pmatrix}_{2\times 2} \qquad B = \begin{pmatrix} 2 & 4 \\ 3 & 1 \end{pmatrix}_{2\times 2} \qquad C = \begin{pmatrix} 5 & -8 & 23 & 4 \end{pmatrix}_{1\times 4}$$

NULA Y CUADRADA   CUADRADA   FILA

DIAGONAL MENOR

$$D = \begin{pmatrix} 12 \\ 4 \\ 9 \end{pmatrix}_{3\times 1} \qquad\qquad E = \begin{pmatrix} 4 & 8 & -5 \\ 0 & 0 & 2 \\ 13 & -4 & -10 \end{pmatrix}_{3\times 3}$$

COLUMNA   CUADRADA

DIAGONAL MAYOR

EQUIVALENT MATRICES (A~B)

They are said to be equivalent when they have the same answers.

First form. Interchange between two rows or columns.

$$\boxed{F1 \longleftrightarrow F2}$$

$$\begin{pmatrix} 3 & -2 & 5 \\ 4 & -6 & 8 \\ -7 & 0 & 4 \end{pmatrix} \sim \begin{pmatrix} -7 & 0 & 4 \\ 4 & 4 & 3 \\ 3 & -2 & 5 \end{pmatrix}$$

Second way. Multiply a row or columns by a number other than zero.

$$\boxed{F3 \longleftrightarrow 4F3}$$

$$\begin{pmatrix} 3 & -2 & 5 \\ 4 & -6 & 8 \\ -7 & 0 & 4 \end{pmatrix} \sim \begin{pmatrix} -7 & 0 & 4 \\ 4 & 4 & 3 \\ -28 & 0 & 16 \end{pmatrix}$$

Third way. Add to a row with another row (columns).

$$\boxed{F2 \longleftrightarrow F2 + F1}$$

$$\begin{pmatrix} 3 & -2 & 5 \\ 4 & -6 & 8 \\ -7 & 0 & 4 \end{pmatrix} \sim \begin{pmatrix} 3 & -2 & 5 \\ 7 & -8 & 13 \\ -7 & 0 & 4 \end{pmatrix}$$

Fourth way. Delete or add a NULL row or column.

$$\begin{pmatrix} 3 & -2 & 5 \\ 4 & -6 & 8 \\ -7 & 0 & 4 \end{pmatrix} \sim \begin{pmatrix} 3 & -2 & 5 \\ 4 & -6 & 8 \\ -7 & 0 & 4 \\ \mathbf{0} & \mathbf{0} & \mathbf{0} \end{pmatrix}$$

Example.

$$\begin{pmatrix} 3 & -2 & 5 \\ 4 & 0 & -2 \\ -3 & 9 & 1 \end{pmatrix} \qquad F3 + F2 \begin{pmatrix} 3 & -2 & 5 \\ 4 & -6 & 8 \\ \mathbf{1} & \mathbf{9} & \mathbf{-1} \end{pmatrix}$$

$$2(F2)\begin{pmatrix} 3 & -2 & 5 \\ \mathbf{8} & \mathbf{0} & \mathbf{-4} \\ 1 & 9 & -1 \end{pmatrix} \qquad F1+ F3 \begin{pmatrix} \mathbf{4} & \mathbf{7} & \mathbf{4} \\ 8 & 0 & -4 \\ 1 & 9 & -1 \end{pmatrix}$$

TRANSPOSED MATRIX AND ITS CHARACTERISTICS

It is called Matrix Transpose of **A** and is designated **A**^{t} to the matrix that is obtained by changing the rows by the columns in orderly order.

Examples, if you have a 3x3 matrix you get a 3x3 matrix transpose.

$$A = \begin{pmatrix} 5 & -2 & 8 \\ 7 & 1 & 0 \\ -4 & -7 & -6 \end{pmatrix}_{3x3}$$

$$A^t = \begin{pmatrix} 5 & 7 & -4 \\ -2 & 1 & -7 \\ 8 & 0 & -6 \end{pmatrix}_{3x3}$$

And for a 2x4 matrix we obtain a 4x2 matrix transpose.

$$B = \begin{pmatrix} 1 & 0 & -5 & 6 \\ 2 & 8 & 4 & -9 \end{pmatrix}_{2x4}$$

$$B^t = \begin{pmatrix} 1 & 2 \\ 0 & 8 \\ -5 & 4 \\ 6 & -9 \end{pmatrix}_{4x2}$$

A Square Matrix is called SYMMETRIC if it is equal to its transpose.

$$M = \begin{pmatrix} 2 & 4 & -8 \\ 4 & -3 & 5 \\ -8 & 5 & 0 \end{pmatrix}_{3 \times 3}$$

$$M^t = \begin{pmatrix} 2 & 4 & -8 \\ 4 & -3 & 5 \\ -8 & 5 & 0 \end{pmatrix}_{3 \times 3}$$

A *property of the Transpose* is that the transpose of the same transpose of a matrix is equal to the original matrix.

$$\boxed{(A^t)^t = A} \qquad A = \begin{pmatrix} 6 & 4 \\ -2 & 8 \\ 3 & 1 \end{pmatrix}_{3 \times 2}$$

$$A^t = \begin{pmatrix} 6 & -2 & 3 \\ 4 & 8 & -1 \end{pmatrix}_{2 \times 3} \qquad (A^t)^t = \begin{pmatrix} 6 & 4 \\ -2 & 8 \\ 3 & 1 \end{pmatrix}_{3 \times 2}$$

Examples:

$$M = \begin{pmatrix} -1 & 0 & 5 \\ 8 & 2 & 3 \end{pmatrix}_{2 \times 3} \qquad M^t = \begin{pmatrix} -1 & 8 \\ 0 & 2 \\ 5 & 3 \end{pmatrix}_{3 \times 2}$$

$$N = \begin{pmatrix} 0 & 1 & 4 \\ -2 & 8 & 3 \\ 15 & -4 & 0 \end{pmatrix}_{3 \times 3} \qquad N^t = \begin{pmatrix} 0 & -2 & 15 \\ 1 & 8 & -4 \\ 4 & 3 & 0 \end{pmatrix}_{3 \times 3}$$

$$D = \begin{pmatrix} 0 & 1 \\ 2 & 3 \end{pmatrix}_{2 \times 2} \qquad D^t = \begin{pmatrix} 0 & 2 \\ 1 & 3 \end{pmatrix}_{2 \times 2}$$

MATRIX OPERATIONS

SUM OF MATRICES (A+B)

The addition of two matrices can only be performed between matrices of the **SAME ORDER**.

$$A = \begin{pmatrix} -3 & 0 \\ 2 & 5 \\ 8 & -7 \end{pmatrix}_{3x2} \qquad B = \begin{pmatrix} 7 & -5 \\ 4 & -2 \\ 1 & -4 \end{pmatrix}_{3x2}$$

$$A + B = \begin{pmatrix} -3 + 7 & 0 + (-5) \\ 2 + 4 & 5 + (-2) \\ 8 + 1 & -7 + (-4) \end{pmatrix} \qquad A + B = \begin{pmatrix} 4 & -5 \\ 6 & 3 \\ 9 & -11 \end{pmatrix}$$

Example.

$$A = \begin{pmatrix} -5 & 10 & 0 \\ 2 & -14 & 9 \\ -6 & 3 & 8 \end{pmatrix} \quad B = \begin{pmatrix} -8 & 3 & 4 \\ -9 & 7 & -5 \\ 15 & -3 & 0 \end{pmatrix} \quad C = \begin{pmatrix} 5 & 0 & 2 \\ -4 & 6 & 10 \\ 0 & -5 & 1 \end{pmatrix}$$

$$A + B = \begin{pmatrix} -13 & 13 & 4 \\ -7 & -7 & 4 \\ 9 & 0 & 8 \end{pmatrix} \qquad B + C = \begin{pmatrix} -3 & 3 & 6 \\ -13 & 13 & 5 \\ 15 & -8 & 1 \end{pmatrix}$$

$$A + C = \begin{pmatrix} 0 & 10 & 2 \\ -2 & -8 & 19 \\ -6 & -2 & 9 \end{pmatrix} \qquad A + B + C = \begin{pmatrix} -8 & 13 & 6 \\ -11 & -1 & 14 \\ 9 & -5 & 9 \end{pmatrix}$$

SUBTRACTION OF MATRICES (A-B)

Subtraction of two matrices can only be performed between matrices of the **SAME ORDER**.

$$A = \begin{pmatrix} -3 & 0 \\ 2 & 5 \\ 8 & -7 \end{pmatrix}_{3x2} \qquad B = \begin{pmatrix} 7 & -5 \\ 4 & -2 \\ 1 & -4 \end{pmatrix}_{3x2}$$

$$A - B = \begin{pmatrix} -3 - 7 & 0 - (-5) \\ 2 - 4 & 5 - (-2) \\ 8 - 1 & -7 - (-4) \end{pmatrix} \qquad A - B = \begin{pmatrix} -10 & 5 \\ -2 & 7 \\ 7 & -3 \end{pmatrix}$$

Another way to perform the subtraction operations is to change the signs of all the elements of the matrix B.

$$\boxed{A - B = A + (-B)}$$

$$A = \begin{pmatrix} -3 & 0 \\ 2 & 5 \\ 8 & -7 \end{pmatrix}_{3x2} \qquad B = \begin{pmatrix} 7 & -5 \\ 4 & -2 \\ 1 & -4 \end{pmatrix}_{3x2}$$

$$A - B = \begin{pmatrix} -3 - 7 & 0 - (-5) \\ 2 - 4 & 5 - (-2) \\ 8 - 1 & -7 - (-4) \end{pmatrix} \qquad A - B = \begin{pmatrix} -10 & 5 \\ -2 & 7 \\ 7 & -3 \end{pmatrix}$$

Example.

$$A = \begin{pmatrix} 0 & -8 & 9 \\ 14 & -3 & 5 \\ 20 & 0 & -15 \end{pmatrix} \qquad B = \begin{pmatrix} 4 & -2 & 6 \\ 7 & 5 & 12 \\ 0 & -2 & -8 \end{pmatrix}$$

$$-A = \begin{pmatrix} 0 & 8 & -9 \\ -14 & 3 & -5 \\ -20 & 0 & 15 \end{pmatrix} \qquad -B = \begin{pmatrix} -4 & 2 & -6 \\ -7 & -5 & -12 \\ 0 & 2 & 8 \end{pmatrix}$$

$$A - B = \begin{pmatrix} -4 & -6 & 3 \\ 7 & -8 & -7 \\ 20 & 2 & -7 \end{pmatrix} \qquad B - A = \begin{pmatrix} 4 & 6 & -3 \\ -7 & 8 & 7 \\ -20 & -2 & 7 \end{pmatrix}$$

PRODUCT OF A MATRIX BY A SCALAR OR REAL (K - A)

Given a matrix $A = (a)_{ijmxn}$ and a real number K, the product $K-A$ is made by multiplying **ALL** the elements of A by K, resulting in another matrix of equal size.

$$4 \times \begin{pmatrix} -2 & 0 \\ 3 & -5 \\ 12 & -8 \end{pmatrix}_{3x2} = \begin{pmatrix} -8 & 0 \\ 12 & 20 \\ 48 & -32 \end{pmatrix}$$

Examples.

$$-5 \times \begin{pmatrix} -7 & 15 \\ 0 & -9 \\ 3 & -2 \end{pmatrix} = \begin{pmatrix} 35 & -75 \\ 0 & 45 \\ -15 & 10 \end{pmatrix}$$

$$(3/2) \times \begin{pmatrix} 2 & -4 \\ 1 & -8 \end{pmatrix} = \begin{pmatrix} 3 & -6 \\ 3/2 & -12 \end{pmatrix}$$

$$3 \times \begin{pmatrix} 4 & 0 & 5 \\ -3 & 8 & -15 \\ -12 & 0 & 6 \end{pmatrix} = \begin{pmatrix} 12 & 0 & 15 \\ -9 & 24 & -45 \\ -36 & 0 & 18 \end{pmatrix}$$

$$(-4/3) \times \begin{pmatrix} 6 & -2 & 12 \\ 0 & -1 & 9 \\ 5 & 3 & -8 \end{pmatrix} = \begin{pmatrix} -8 & 2.6 & -16 \\ -9 & 1.3 & -12 \\ -6.6 & -4 & 10.6 \end{pmatrix}$$

MATRIX MULTIPLICATION (A - B)

For the ***multiplication of two matrices*** it is necessary that the **NUMBER OF COLUMNS** of the **FIRST MATRIX** is **EQUAL** to the **NUMBER OF ROWS** of the **SECOND MATRIX**.

Matrix A = Matrix B

Rows x $\boxed{\textit{Columns = Rows}}$ *x Columns*

$$A = \begin{pmatrix} 5 & 3 & -4 & -2 \\ 8 & -1 & 0 & -3 \end{pmatrix}_{2x4} \qquad B = \begin{pmatrix} 1 & 4 & 0 \\ -5 & 3 & 7 \\ 0 & -9 & 5 \\ 5 & 1 & 4 \end{pmatrix}_{4x3}$$

$$(2) \times \boxed{4} \checkmark \boxed{4} \times (3)$$

$$C = \begin{pmatrix} C_{11} & C_{12} & C_{13} \\ C_{21} & C_{22} & C_{23} \end{pmatrix}_{2x3}$$

And the following results are obtained:

$$A = \begin{pmatrix} 5 & 3 & -4 & -2 \\ 8 & -1 & 0 & -3 \end{pmatrix}_{2x4} \qquad B = \begin{pmatrix} 1 & 4 & 0 \\ -5 & 3 & 7 \\ 0 & -9 & 5 \\ 5 & 1 & 4 \end{pmatrix}_{4x3}$$

$$\boxed{A \times B = C}$$

* *La multiplicación se hace Filas x Columnas*

$$C_{11} = 5 - 15 + 0 - 10 = -20$$
$$C_{12} = 20 + 9 + 36 - 2 = 63$$
$$C_{13} = 21 - 20 - 8 = -7$$
$$C_{21} = 8 + 5 - 15 = -2$$
$$C_{22} = 32 - 3 - 3 = 26$$
$$C_{23} = -7 - 12 = -19$$

$$C = \begin{pmatrix} -20 & 63 & -7 \\ -2 & 26 & -19 \end{pmatrix}_{2x3}$$

Example of 2x2

$$A = \begin{pmatrix} -5 & 3 \\ 4 & 7 \end{pmatrix}_{2x2} \qquad B = \begin{pmatrix} 9 & 0 \\ 2 & -5 \end{pmatrix}_{2x2}$$

$$\boxed{A \times B = C}$$

$C_{11} = -45 + 6 = -39$
$C_{12} = 0 - 15 = -15$
$C_{21} = 36 + 14 = 50$
$C_{22} = 0 - 35 = -35$

$$C = \begin{pmatrix} -39 & -15 \\ 50 & -35 \end{pmatrix}_{2x2}$$

Example of 3x3

$$A = \begin{pmatrix} 0 & -7 & 3 \\ 2 & 4 & -1 \\ 12 & 7 & -6 \end{pmatrix}_{3x3} \qquad B = \begin{pmatrix} 5 & 4 & -3 \\ 0 & -6 & 10 \\ -2 & 8 & 11 \end{pmatrix}_{3x3}$$

$$\boxed{A \times B = C}$$

$C_{11} = -6 \qquad\qquad = -6$
$C_{12} = 42 + 24 \qquad = 66$
$C_{13} = -70 + 33 \qquad = -37$
$C_{21} = 10 + 2 \qquad\quad = 12$
$C_{22} = 8 - 24 - 8 \qquad = -24$
$C_{23} = -6 + 40 - 11 = 23$
$C_{31} = 60 + 12 \qquad\quad = 72$
$C_{32} = 48 - 42 - 48 = -42$
$C_{33} = -36 + 70 - 66 = -32$

$$C = \begin{pmatrix} -6 & 66 & -37 \\ 12 & -24 & 23 \\ 72 & -42 & -32 \end{pmatrix}_{3x3}$$

Examples.

$$A = \begin{pmatrix} 0 & -7 \\ 2 & 4 \\ 12 & 7 \end{pmatrix}_{3\times2} \qquad B = \begin{pmatrix} 5 & 4 & -3 \\ 0 & -6 & 10 \end{pmatrix}_{2\times3}$$

$$\boxed{A \times B = C}$$

$C_{11} = -2 - 24 = -26$
$C_{12} = 10 + 27 = 37$
$C_{13} = 0 + 6 = 6$
$C_{21} = -5 - 8 = -13$
$C_{22} = 25 + 9 = 34$
$C_{23} = 0 + 2 = 2$
$C_{31} = 0 + 48 = 48$
$C_{32} = 0 - 54 = -54$
$C_{33} = 0 + 12 = -12$

$$C = \begin{pmatrix} -26 & 37 & 6 \\ -13 & 34 & 2 \\ 48 & -54 & -12 \end{pmatrix}_{3\times3}$$

Example of 2x2

$$A = \begin{pmatrix} 6 & -2 \\ -8 & 10 \end{pmatrix}_{2\times2} \qquad B = \begin{pmatrix} 1 & -3 \\ 2 & 12 \end{pmatrix}_{2\times2}$$

$$\boxed{A \times B = C}$$

$C_{11} = 6 - 4 = 2$
$C_{12} = -18 - 24 = -42$
$C_{21} = -8 + 20 = 12$
$C_{22} = 24 + 120 = 144$

$$C = \begin{pmatrix} 2 & -42 \\ 12 & 144 \end{pmatrix}_{2\times2}$$

Example of 3x3

$$A = \begin{pmatrix} -2 & 0 & 1 \\ 8 & 4 & 3 \\ 3 & -1 & 0 \end{pmatrix}_{3\times3} \qquad B = \begin{pmatrix} 5 & -2 & 4 \\ 0 & 8 & 3 \\ 1 & 0 & -1 \end{pmatrix}_{3\times3}$$

$$\boxed{A \times B = C}$$

$C_{11} = -10 + 0 + 1 = -9$

$C_{12} = 4 + 0 + 0 = 4$

$C_{13} = -8 + 0 - 1 = -9$

$C_{21} = 40 + 0 + 3 = 43$

$C_{22} = -16 + 32 + 0 = 16$

$C_{23} = 32 + 12 - 3 = 41$

$C_{31} = 15 + 0 + 0 = 15$

$C_{32} = -6 - 8 + 0 = -14$

$C_{33} = 12 - 3 + 0 = -9$

$$C = \begin{pmatrix} -9 & 4 & -9 \\ 43 & 16 & 41 \\ 15 & -14 & 9 \end{pmatrix}_{3\times3}$$

CASES WHEN MATRICES CANNOT BE MULTIPLIED

Not all matrices can be multiplied. Recall that the *first condition* for *multiplying two matrices* is that the *number of columns of the first matrix* must match or be *EQUAL* to the *number of rows of the second matrix.*

Therefore, the following cases of matrix multiplication cannot be performed, because the first matrix has 3 columns and, on the other hand, the second matrix has 2 rows:

$$\begin{pmatrix} 1 & 3 & -2 \\ 4 & 0 & 5 \end{pmatrix} \cdot \begin{pmatrix} 2 & 1 \\ 3 & -1 \end{pmatrix} \quad \longleftarrow \ \times$$

But if we reverse the order, they can be multiplied. Because the first matrix has two columns and the second matrix has two rows:

$$\begin{pmatrix} 2 & 1 \\ 3 & -1 \end{pmatrix} \cdot \begin{pmatrix} 1 & 3 & -2 \\ 4 & 0 & 5 \end{pmatrix} = \begin{pmatrix} 2 \cdot 1 + 1 \cdot 4 & 2 \cdot 3 + 1 \cdot 0 & 2 \cdot (-2) + 1 \cdot 5 \\ 3 \cdot 1 + (-1) \cdot 4 & 3 \cdot 3 + (-1) \cdot 0 & 3 \cdot (-2) + (-1) \cdot 5 \end{pmatrix}$$

We would have as a result:

$$= \begin{pmatrix} 6 & 6 & 1 \\ -1 & 9 & -11 \end{pmatrix}$$

PROPERTIES OF MATRIX MULTIPLICATION

This type of matrix operation has the following characteristics:

- Matrix multiplication is **associative**:

$$(A \cdot B) \cdot C = A \cdot (B \cdot C)$$

- Matrix multiplication also has the **distributive** property:

$$A \cdot (B + C) = A \cdot B + A \cdot C$$

- The product of matrices **is not commutative**:

$$A \cdot B \neq B \cdot A$$

For example, the following matrix multiplication gives a result:

$$\begin{pmatrix} 1 & -1 \\ 2 & 3 \end{pmatrix} \cdot \begin{pmatrix} -2 & 5 \\ 0 & 1 \end{pmatrix} = \begin{pmatrix} 1 \cdot (-2) + (-1) \cdot 0 & 1 \cdot 5 + (-1) \cdot 1 \\ 2 \cdot (-2) + 3 \cdot 0 & 2 \cdot 5 + 3 \cdot 1 \end{pmatrix}$$

Resulting in:

$$= \begin{pmatrix} -2 & 4 \\ -4 & 13 \end{pmatrix}$$

But the result of the product is different if we reverse the order of multiplication of the matrices:

$$\begin{pmatrix} -2 & 5 \\ 0 & 1 \end{pmatrix} \cdot \begin{pmatrix} 1 & -1 \\ 2 & 3 \end{pmatrix} = \begin{pmatrix} -2 \cdot 1 + 5 \cdot 2 & -2 \cdot (-1) + 5 \cdot 3 \\ 0 \cdot 1 + 1 \cdot 2 & 0 \cdot (-1) + 1 \cdot 3 \end{pmatrix}$$

The following result was obtained:

$$= \begin{pmatrix} 8 & 17 \\ 2 & 3 \end{pmatrix}$$

As we realize, in this **commutative property** the order of the factors **DOES** affect the result.

- Moreover, any matrix multiplied by the identity matrix results in the same matrix. This is called the **multiplicative identity property**:

$$A \cdot I = A$$

$$I \cdot A = A$$

For example:

$$\begin{pmatrix} 2 & 7 \\ -6 & 5 \end{pmatrix} \cdot \begin{pmatrix} 1 & 0 \\ 0 & 1 \end{pmatrix} = \begin{pmatrix} 2 & 7 \\ -6 & 5 \end{pmatrix}$$

- And finally, any matrix multiplied by the null matrix is equal to the null matrix. This is called the **multiplicative property of zero**:

$$A \cdot 0 = 0$$

$$0 \cdot A = 0$$

For example:

$$\begin{pmatrix} 6 & -4 \\ 3 & 8 \end{pmatrix} \cdot \begin{pmatrix} 0 & 0 \\ 0 & 0 \end{pmatrix} = \begin{pmatrix} \mathbf{0} & \mathbf{0} \\ \mathbf{0} & \mathbf{0} \end{pmatrix}$$

COMPLEMENTARY MINOR, ADJUNCT AND ADJUNCT MATRIX

THE LEAST COMPLEMENTARY OF A MATRIX

Given a **square matrix A** of **order n**, i.e. 2x2, 3x3, etc... the least complementary of an element of $A(a_{ij})$ is the determinant obtained by deleting the row (**i**) and column (**j**) in which a_{ij} is located and is represented by M_{ij} .

Example.

The minor complementary of M_{11} .

$$A = \begin{pmatrix} -5 & 0 & 4 \\ 8 & -3 & 2 \\ 5 & 1 & 7 \end{pmatrix}$$

EL *MENOR COMPLEMENTARIO* DE **-5**

$$M_{11} = \begin{vmatrix} -3 & 2 \\ 1 & 7 \end{vmatrix} = -21 - 2 = \boxed{-23}$$

* APLICANDO DETERMINANTES

And that of M_{23} .

$$M_{23} = \begin{vmatrix} -5 & 0 \\ 5 & 1 \end{vmatrix} = -5 - 0 = \boxed{-5}$$

ATTACHMENT OF A MATRIX

The Attachment (A_{ij}) of a Matrix is obtained with the following formula:

$$A_{ij} = (-1)^{i+j} \times M_{ij}$$

Continuing with the first complementary Minor (M_{11}), the Adjunct (A_{11}) would be obtained as follows:

$$M_{11} = -23 \qquad A = \begin{pmatrix} -5 & 0 & 4 \\ 8 & -3 & 2 \\ 5 & 1 & 7 \end{pmatrix}$$

$$A_{ij} = (-1)^{i+j} \, M_{ij}$$
$$A_{11} = (-1)^{1+1} \times (-23)$$
$$A_{11} = (-1)^{2} \times (-23)$$
$$A_{11} = (1) \times (-23)$$
$$A_{11} = \mathbf{-23}$$

But it can also be calculated by creating a ***matrix of signs*** of equal order, putting the signs (+ -) and intercalating them by rows, starting with the positive (+), to proceed to multiply the complementary minor obtained by the sign of where the calculated value is located, example.

$$\begin{pmatrix} + & - & + \\ - & + & - \\ + & - & + \end{pmatrix}$$

$$M_{11} = +\!-23$$
$$= -23$$

Note: It will not always be the result of the same sign as the calculated Attachment.

For example, the complementary Minor of $M_{23} = -5$, its Adjunct would be $A_{23} = \mathbf{5}$, i.e. of opposite sign.

To know the reason for the result, the calculation is applied with the same formula.

$$A_{ij} = (-1)^{i+j}\, M_{ij}$$

$$M_{23} = 5 \qquad A = \begin{pmatrix} -5 & 0 & 4 \\ 8 & -3 & 2 \\ 5 & 1 & 7 \end{pmatrix}$$

$$A_{ij} = (-1)^{i+j}\, M_{ij}$$
$$A_{23} = (-1)^{2+3} \times (-5)$$
$$A_{23} = (-1)^{5} \times (-5)$$
$$A_{23} = (-1) \times (-5)$$
$$A_{23} = 5$$

With the *sign matrix*, the same result is obtained.

$$\begin{pmatrix} + & - & + \\ - & + & - \\ + & - & + \end{pmatrix}$$

$$M_{23} = - -5$$
$$= 5$$

Examples.

$$A = \begin{pmatrix} -5 & 0 & 4 \\ 8 & -3 & 2 \\ 5 & 1 & 7 \end{pmatrix} \qquad\qquad \begin{pmatrix} + & - & + \\ - & + & - \\ + & - & + \end{pmatrix}$$

$$M_{22} = \begin{vmatrix} -5 & 4 \\ 5 & 7 \end{vmatrix} = -35 - 20 = \boxed{-55} \qquad A_{22} = +-55 = \mathbf{-55}$$

$$M_{31} = \begin{vmatrix} 0 & 4 \\ -3 & 2 \end{vmatrix} = 0 - (-12) = \boxed{12} \qquad A_{31} = ++12 = \mathbf{12}$$

$$M_{32} = \begin{vmatrix} 8 & 2 \\ 5 & 7 \end{vmatrix} = 56 - 10 = \boxed{46} \qquad A_{32} = -+46 = \mathbf{-46}$$

ADJUNCT OR COFACTOR MATRIX

Given a square Matrix (**A**), its Adjoint or cofactor Matrix (**Adj(A)**) is the result of substituting each term (**a**$_{ij}$) of A by the cofactor (**a**$_{ij}$) of (**A**).

$$a_{ij} = (-1)^{i+j} \times M_{ij}$$

a$_{ij}$ = ADJUNTO **ij**

M$_{ij}$ = MENOR COMPLEMENTARIO **ij**

Example.

$$B = \begin{pmatrix} -3 & 2 & 0 \\ 1 & -1 & 2 \\ -2 & 1 & 3 \end{pmatrix}$$

Based on the fact that the order of the signs is interchanging:

$$B = \begin{pmatrix} -3 & 2 & 0 \\ 1 & -1 & 2 \\ -2 & 1 & 3 \end{pmatrix} \qquad \begin{pmatrix} + & - & + \\ - & + & - \\ + & - & + \end{pmatrix}$$

Calculation of the cofactors of **a**$_{11}$, **a**$_{12}$ **and a**$_{13}$:

$$a_{11} = + \begin{vmatrix} -1 & 2 \\ 1 & 3 \end{vmatrix} = +(-3 - 2) = -5$$

$$a_{12} = - \begin{vmatrix} 1 & 2 \\ -2 & 3 \end{vmatrix} = -(3 + 4) = -7$$

$$a_{13} = + \begin{vmatrix} 1 & -1 \\ -2 & 1 \end{vmatrix} = +(1 - 2) = -1$$

Calculation of the cofactors of **a**$_{21}$, **a**$_{22}$ **and a**$_{23}$:

$$a_{21} = - \begin{vmatrix} 2 & 0 \\ 1 & 3 \end{vmatrix} = -(6 - 0) = \boxed{-6}$$

$$a_{22} = + \begin{vmatrix} -3 & 0 \\ -2 & 3 \end{vmatrix} = +(-9 - 0) = \boxed{-9}$$

$$a_{23} = - \begin{vmatrix} -3 & 2 \\ -2 & 1 \end{vmatrix} = -(-3 + 4) = \boxed{-1}$$

Calculation of the cofactors of **a_{31}**, **a_{32} and a_{33}** :

$$a_{31} = + \begin{vmatrix} 2 & 0 \\ -1 & 2 \end{vmatrix} = +(4 - 0) = \boxed{4}$$

$$a_{32} = - \begin{vmatrix} -3 & 0 \\ 1 & 2 \end{vmatrix} = -(-6 - 0) = \boxed{6}$$

$$a_{33} = + \begin{vmatrix} -3 & 2 \\ 1 & -1 \end{vmatrix} = +(3 - 2) = \boxed{1}$$

Result of the Adjunct or Cofactor Matrix.

$$Adj\,(B) = \begin{pmatrix} -5 & -7 & -1 \\ -6 & -9 & -1 \\ 4 & 6 & 1 \end{pmatrix}$$

Exercise.

Matrix C

$$C = \begin{pmatrix} 2 & -2 & 2 \\ 2 & 1 & 0 \\ 3 & -2 & 2 \end{pmatrix}$$

Adjunct Matrix (**Adj(C)**)

$$Adj\,(C) = \begin{pmatrix} 2 & -4 & -7 \\ 0 & -2 & -2 \\ -2 & 4 & 6 \end{pmatrix}$$

THE INVERSE MATRIX

The ***inverse matrix*** of a matrix is equal to the adjoint matrix of its transpose matrix, divided by its determinant, provided this is not zero.

Technical definition

*An **inverse matrix** is the linear transformation of a matrix by multiplying the inverse of the determinant of the matrix by the adjoining transposed matrix.*

In other words, an inverse matrix is the multiplication of the inverse of the determinant by the transposed adjoint matrix.

Theorem:

"Let **A** be a regular matrix of dimension **n**, then it has only one inverse matrix."

Since the inverse matrix of a matrix **A** is unique, we can give it its own name: $\mathbf{A}^{-1}$.

PROPERTIES OF THE INVERSE MATRIX

Let **A** and **B** be two regular matrices of dimension **n**, then:

- *The inverse matrix of* **A**, $\mathbf{A}^{-1}$, *is regular and its inverse is* **A***:*

$$(\mathbf{A}^{-1})^{-1} = \mathbf{A}$$

- *Inverse of the product of matrices:*

$$(\mathbf{AB})^{-1} = \mathbf{B}^{-1}\mathbf{A}^{-1}$$

- *Inverse of the transpose matrix:*

$$(\mathbf{A}^{T})^{-1} = (\mathbf{A}^{-1})^{T}$$

SOLUTION METHODS TO OBTAIN THE INVERSE MATRIX

INVERSE MATRIX BY THE GAUSS-JORDAN METHOD

To obtain the inverse matrix of a base matrix, an identity matrix must be added.

$$\begin{pmatrix} 2 & 3 & 4 \\ 4 & 3 & 2 \\ 0 & 1 & 0 \end{pmatrix} \text{ Matriz base}$$

$$\left(\begin{array}{ccc|ccc} 2 & 3 & 4 & 1 & 0 & 0 \\ 4 & 3 & 2 & 0 & 1 & 0 \\ 0 & 1 & 0 & 0 & 0 & 1 \end{array}\right)$$

Matriz base | Matriz identidad

Example of 2x2 by the **Gauss-Jordan Method.**

$$A = \begin{pmatrix} 1 & 3 \\ 2 & 4 \end{pmatrix}$$

We add the Identity Matrix to start the method by first making zero the value of a_{21} **(2)** and then with the results the value of a_{12} **(3)** of the Base Matrix.

$$\left(\begin{array}{cc|cc} 1 & ③ & 1 & 0 \\ ② & 4 & 0 & 1 \end{array}\right)$$

Solution steps

$$\begin{matrix} 4F_1 - 3F_2 \\ F_2 - 2F_1 \end{matrix} \left(\begin{array}{cc|cc} -2 & 0 & 4 & -3 \\ 0 & -2 & -2 & 1 \end{array}\right)$$

$$\left(\begin{array}{cc|cc} -2 & 0 & 4 & -3 \\ 0 & -2 & -2 & 1 \end{array}\right) \begin{matrix} F_1/-2 \\ F_2/-2 \end{matrix}$$

$$\begin{matrix} F_1/-2 \\ F_2/-2 \end{matrix} \left(\begin{array}{cc|cc} 1 & 0 & -2 & 3/2 \\ 0 & 1 & 1 & -1/2 \end{array}\right)$$

So the **inverse Matrix** of A is:

$$A^{-1} = \begin{pmatrix} -2 & 3/2 \\ 1 & -1/2 \end{pmatrix}$$

The check is performed by **MULTIPLYING** the Base Matrix (**A**) by the Inverse Matrix (**A** x **A^{-1}**), from which we must obtain the Identity Matrix (**I**).

$$(\textit{Filas de } \mathbf{A} \text{ x } \textit{Columnas de } \mathbf{A}^{-1})$$

$$A = \begin{pmatrix} 1 & 3 \\ 2 & 4 \end{pmatrix} \qquad A^{-1} = \begin{pmatrix} -2 & 3/2 \\ 1 & -1/2 \end{pmatrix}$$

$i_{11} = -2 + 3 = 1$
$i_{12} = 3/2 - 3/2 = 0$
$i_{21} = -4 + 4 = 0$
$C_{22} = 3 - 2 = 1$

$$\begin{pmatrix} 1 & 0 \\ 0 & 1 \end{pmatrix} \quad \textit{Matriz identidad (I)}$$

Example 2.

$$B = \begin{pmatrix} 2 & 6 \\ 1 & 4 \end{pmatrix} \qquad \left(\begin{array}{cc|cc} 2 & ⑥ & 1 & 0 \\ ① & 4 & 0 & 1 \end{array}\right)$$

$$\begin{array}{c} 4F_1 - 6F_2 \\ F_1 - 2F_2 \end{array} \left(\begin{array}{cc|cc} 2 & 0 & 4 & -6 \\ 0 & -2 & 1 & -2 \end{array}\right) \qquad \begin{array}{c} F_1/2 \\ F_2/-2 \end{array} \left(\begin{array}{cc|cc} 1 & 0 & 2 & -3 \\ 0 & 1 & -1/2 & 1 \end{array}\right)$$

$$\underset{I}{} \qquad \underset{B^{-1}}{}$$

$$B^{-1} = \begin{pmatrix} 2 & -3 \\ -1/2 & 1 \end{pmatrix}$$

INVERSE MATRIX BY THE INVERSE METHOD

Example 1 of 3x3

Having a Base Matrix (**A**), the Identity Matrix is added to start the procedure.

Matriz Aumentada

$$A = \begin{pmatrix} 2 & 2 & -1 \\ 1 & -3 & -2 \\ 3 & 4 & 1 \end{pmatrix} \qquad \left(\begin{array}{ccc|ccc} 2 & 2 & -1 & 1 & 0 & 0 \\ 1 & -3 & -2 & 0 & 1 & 0 \\ 3 & 4 & 1 & 0 & 0 & 1 \end{array}\right)$$

Procedure.

To start the first iteration of the method we select the value of **(2)** of the first row and first column of the Base Matrix as our *first pivot* of the determinants that we are going to calculate by converting it to the unit (1).

$$\left(\begin{array}{ccc|ccc} \boxed{2} & 2 & -1 & 1 & 0 & 0 \\ 1 & -3 & -2 & 0 & 1 & 0 \\ 3 & 4 & 1 & 0 & 0 & 1 \end{array}\right) \qquad \left(\begin{array}{ccc|ccc} \boxed{1} & 2 & -1 & 1 & 0 & 0 \\ 1 & -3 & -2 & 0 & 1 & 0 \\ 3 & 4 & 1 & 0 & 0 & 1 \end{array}\right)$$

The values of the pivot rule pass through the same and the values below the pivot become zeros.

$$\left(\begin{array}{ccc|ccc} 1 & 2 & -1 & 1 & 0 & 0 \\ 0 & & & & & \\ 0 & & & & & \end{array}\right)$$

And with the pivot, determinants are applied, obtaining the following results:

$$\left(\begin{array}{ccc|ccc} 1 & 2 & -1 & 1 & 0 & 0 \\ 0 & -8 & -3 & -1 & 2 & 0 \\ 0 & 2 & 5 & -3 & 0 & 2 \end{array}\right)$$

And we continue with the second iteration with the **second pivot (-8)**, but now all determinant will be divided by the previous pivot **(2)**, obtaining the following results.

$$\left(\begin{array}{ccc|ccc} 2 & 2 & -1 & 1 & 0 & 0 \\ 0 & \boxed{-8} & -3 & -1 & 2 & 0 \\ 0 & 2 & 5 & -3 & 0 & 2 \end{array}\right) \xrightarrow{\div 2} \left(\begin{array}{ccc|ccc} 1 & 0 & 7 & -3 & -2 & 0 \\ 0 & 1 & -3 & -1 & 2 & 0 \\ 0 & 0 & -17 & 13 & -2 & -8 \end{array}\right)$$

Third iteration.

$$\left(\begin{array}{ccc|ccc} 1 & 0 & 7 & -3 & -2 & 0 \\ 0 & 1 & -3 & 1 & 2 & 0 \\ 0 & 0 & \boxed{-17} & 13 & -2 & -8 \end{array}\right) \quad \textbf{\textit{Third pivot (-17)}}$$

And the determinants are divided by the previous pivot **(-8)**, obtaining as last determinant value **(-17)** and the following results.

$$\left(\begin{array}{ccc|ccc} 1 & 0 & 7 & -3 & -2 & 0 \\ 0 & 1 & -3 & 1 & 2 & 0 \\ 0 & 0 & -17 & 13 & -2 & -8 \end{array}\right) \xrightarrow{\div 8} \left(\begin{array}{ccc|ccc} 1 & 0 & 0 & 5 & -6 & -7 \\ 0 & 1 & 0 & -7 & 5 & 3 \\ 0 & 0 & 1 & 13 & -2 & -8 \end{array}\right)$$

Finally, the inverse matrix **(A^{-1})** is obtained by multiplying the value of the last determinant **(1/-17)** by each element of the adjoining matrix **(Adj(A))**.

$$\boxed{A^{-1} = \frac{1}{|A|} \times |Adj(A)|} = = \left(\begin{array}{ccc} 5 & -6 & -7 \\ -7 & 5 & 3 \\ 13 & -2 & -8 \end{array}\right)^{\div 17} \left(\begin{array}{ccc} -5/17 & -6/17 & -7/17 \\ -7/17 & 5/17 & 3/17 \\ 13/17 & -2/17 & -8/17 \end{array}\right)$$

Example 2 of 3x3.

$$A = \begin{pmatrix} 2 & 1 & 3 \\ -1 & 2 & 4 \\ 0 & 1 & 3 \end{pmatrix}$$

Adding the Identity Matrix

Matriz Aumentada

$$\left(\begin{array}{ccc|ccc} 2 & 1 & 3 & 1 & 0 & 0 \\ -1 & 2 & 4 & 0 & 1 & 0 \\ 0 & 1 & 3 & 0 & 0 & 1 \end{array} \right)$$

Solution

$$A^{-1} = \begin{pmatrix} 1/2 & 0 & -1/2 \\ 3/4 & 3/2 & -11/4 \\ -1/4 & -1/2 & 5/4 \end{pmatrix}$$

Exercise 2

Matriz Base **Matriz Aumentada**

$$B = \begin{pmatrix} 2 & 3 & 1 \\ 1 & -1 & 2 \\ 0 & 1 & 0 \end{pmatrix} \longrightarrow \left(\begin{array}{ccc|ccc} 2 & 3 & 1 & 1 & 0 & 0 \\ 1 & -1 & 2 & 0 & 1 & 0 \\ 0 & 1 & 0 & 0 & 0 & 1 \end{array} \right)$$

Matriz Inversa

$$\textit{Solución} \quad B^{-1} = \begin{pmatrix} 2/3 & -1/3 & -7/3 \\ 0 & 0 & 1 \\ -1/3 & 2/3 & -5/4 \end{pmatrix}$$

INVERSE MATRIX OF A 2X2 MATRIX BY THE DETERMINANT METHOD.

Suppose we have the following matrix:

$$A = \begin{pmatrix} 1 & 3 \\ 2 & 4 \end{pmatrix}$$

To obtain the ***inverse of the matrix of*** A, we start the method by calculating the value of the determinant (Δ) of the **matrix A**, which is obtained as follows:

$$A = \begin{pmatrix} 1 & 3 \\ 2 & 4 \end{pmatrix} \longrightarrow \quad \Delta = (1 \times 4) - (2 \times 3) = \boxed{-2}$$

Now with respect to the **matrix A** we are going to obtain the ***attached matrix of*** A, making a couple of adjustments, the ***first adjustment*** consists of ***changing the order of the*** elements that form the ***upper diagonal*** or ***dominant diagonal***, that is to say:

$$A = \begin{pmatrix} 1 & 3 \\ 2 & 4 \end{pmatrix} \longrightarrow \quad A = \begin{pmatrix} 4 & 3 \\ 2 & 1 \end{pmatrix}$$

The ***second adjustment*** will be in relation to the elements that make up the ***lower diagonal***, in which ***the sign of each element*** must be ***changed***, finally resulting in the ***attached matrix of*** A:

$$A = \begin{pmatrix} 4 & 3 \\ 2 & 1 \end{pmatrix} \longrightarrow \quad A = \begin{pmatrix} 4 & -3 \\ -2 & 1 \end{pmatrix}$$

And finally ***to obtain the inverse of the matrix of*** A, let's divide each element of the ***adjoint matrix of*** A by the ***value of the determinant*** $(\Delta = -2)$, and thus we obtain the ***inverse of the matrix of*** A.

$$A^{-1} = \frac{A}{\Delta} = \begin{pmatrix} 4 & -3 \\ -2 & 1 \end{pmatrix} \div (-2) = \begin{pmatrix} -2 & 3/2 \\ 1 & -1/2 \end{pmatrix}$$

MATRIX EQUATIONS

WHAT ARE MATRIX EQUATIONS?

Matrix equations are like normal equations, but instead of being made up of numbers, they are made up of matrices. For example:

$$AX = B$$

Therefore, the solution X will also be a matrix.

As you know, matrices cannot be divided. Therefore, you **CANNOT** clear the matrix X by dividing the matrix that multiplied it to the other side of the equation:

$$X = \frac{B}{A}$$

But to clear the matrix X we have to follow a whole procedure. Let's look at the following example of how to clear matrix equations:

$$AX + B = C$$

The elements of each matrix (**A**, **B** and **C**) are as follows:

$$A = \begin{pmatrix} 2 & 1 \\ 4 & 3 \end{pmatrix} \qquad B = \begin{pmatrix} 3 & -1 \\ 0 & 5 \end{pmatrix} \qquad C = \begin{pmatrix} 2 & 1 \\ 6 & -3 \end{pmatrix}$$

The first thing to do is to clear matrix X, so *we subtract matrix B from the other member of the equation*:

$$AX + B = C$$

$$AX = C - B$$

To finish clearing the matrix X, we have to pass the matrix $\mathbf{A}$ to the other member of the equation. However, *we cannot pass it by dividing* as we always did in normal equations, because *matrices cannot be divided*. Instead we must do the following:

> *We have to multiply the two members of the equation by the **inverse of** **the matrix that is multiplying the matrix X** and, in addition, multiply the two members **by the side where that matrix is.***

That is to say that the matrix that multiplies X is $\mathbf{A}$, and it is to its left. Therefore, **we multiply from the left the two members of the equation by the inverse of A (A^{-1})**:

$$AX = C - B$$

$$A^{-1} \cdot AX = A^{-1} \cdot (C - B)$$

Next, let us remember that *a matrix multiplied by its inverse is equal to the matrix identity*, that is; $A^{-1} \times A = I$, therefore, our matrix equation remains:

$$IX = A^{-1} \cdot (C - B)$$

And, furthermore, *any matrix multiplied by the identity matrix results in the same matrix.* Therefore:

$$X = A^{-1} \cdot (C - B)$$

And in this way *we already have the matrix X cleared.* Now we only need to do the matrix operations. So first *we calculate the **inverse** 2×2 **matrix** of* $\mathbf{A}$:

$$A = \begin{pmatrix} 2 & 1 \\ 4 & 3 \end{pmatrix}$$

$$A^{-1} = \frac{1}{|A|} \cdot \left(\mathrm{Adj}(A) \right)^{t}$$

We calculate the *adjoint of the matrix* $\mathbf{A}$:

$$A^{-1} = \frac{1}{2} \cdot \begin{pmatrix} 3 & -4 \\ -1 & 2 \end{pmatrix}^{t}$$

And once the adjoining matrix has been found, we proceed to calculate the **transposed matrix** in order to determine the inverse matrix:

$$A^{-1} = \frac{1}{2} \cdot \begin{pmatrix} 3 & -1 \\ -4 & 2 \end{pmatrix}$$

Now we substitute all the matrices in the expression to calculate the **matrix X**:

$$X = A^{-1} \cdot (C - B)$$

$$X = \begin{pmatrix} \frac{3}{2} & -\frac{1}{2} \\ -2 & 1 \end{pmatrix} \cdot \left(\begin{pmatrix} 2 & 1 \\ 6 & -3 \end{pmatrix} - \begin{pmatrix} 3 & -1 \\ 0 & 5 \end{pmatrix} \right)$$

And we proceed to solve the operations with matrices. First we calculate the parenthesis by doing the matrix subtraction:

$$X = \begin{pmatrix} \frac{3}{2} & -\frac{1}{2} \\ -2 & 1 \end{pmatrix} \begin{pmatrix} -1 & 2 \\ 6 & -8 \end{pmatrix}$$

And, finally, we multiply the matrices:

$$X = \begin{pmatrix} \frac{3}{2} \cdot (-1) + \left(-\frac{1}{2}\right) \cdot 6 & \frac{3}{2} \cdot 2 + \left(-\frac{1}{2}\right) \cdot (-8) \\ -2 \cdot (-1) + 1 \cdot 6 & -2 \cdot 2 + 1 \cdot (-8) \end{pmatrix}$$

$$X = \begin{pmatrix} -\frac{3}{2} - \frac{6}{2} & 3 + 4 \\ 2 + 6 & -4 - 8 \end{pmatrix}$$

$$X = \begin{pmatrix} -\frac{9}{2} & 7 \\ 8 & -12 \end{pmatrix}$$

DETERMINANTS

In mathematics, the determinant is defined as an alternating multilinear form on a vector space. This definition indicates a series of mathematical properties and generalizes the concept of determinant of a matrix making it applicable in numerous fields.

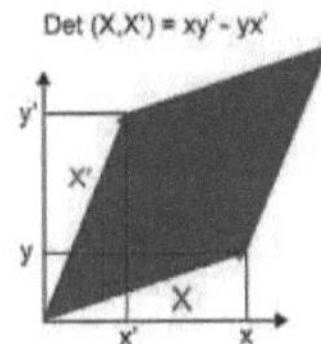

Definition:
If it is a 2 x 2 matrix, the determinant of the matrix A is defined, and is expressed as ***det*(A)** or |**A**|, as the number:

$$det(A) = |A| = \begin{vmatrix} a_{11} & a_{12} \\ a_{21} & a_{22} \end{vmatrix}$$

And it is resolved as follows:

$$det(A) = |A| = \begin{vmatrix} a_{11} & a_{12} \\ a_{21} & a_{22} \end{vmatrix} = a_{11} \cdot a_{22} - a_{12} \cdot a_{21}$$

The way in which a value of a determinant is obtained is through the multiplication of four values formed by ***Rows*** and ***Columns***, where the operation is the subtraction of two multiplications, starting with the cross multiplication of the top value (C_{11}) by the bottom value (C_{22}) in the direction from left to right, minus the bottom value (C_{21}) by the top value (C_{12}) crossed from right to left, that is to say, the operation is the subtraction **of two multiplications, starting** with the cross mul**tiplication of the top value (C)** by the bottom **value (C) by the top value (C)** crossed from right to left, that is to say:

$$\begin{pmatrix} 5 & 8 \\ 7 & 9 \end{pmatrix} - \begin{pmatrix} 5 & 8 \\ 7 & 9 \end{pmatrix}$$

(5 x 9) - (7 x 8)

45 - 56 = -11

PROPERTIES OF DETERMINANTS

First property. *If a matrix has a line of zeros the determinant is zero "0".*

$$\begin{pmatrix} 5 & 0 \\ -3 & 0 \end{pmatrix} \qquad \begin{pmatrix} 8 & -4 & 10 \\ 0 & 0 & 0 \\ 2 & -1 & 6 \end{pmatrix}$$

(5 x 0) - (-3 x 0) = 0   (8 x 0 x 6) - (2 x 0 x 10) = 0

Second property. If a Matrix has two equal lines its determinant is NULL "0".

$$\begin{pmatrix} 6 & 6 \\ 4 & 4 \end{pmatrix} \qquad \begin{pmatrix} 7 & -3 & 10 \\ 5 & 4 & 8 \\ 7 & -3 & 10 \end{pmatrix}$$

(6 x 4) - (4 x 6) = 0 (7 x 4 x 10) - (7 x 4 x 10) = 0

24 - 24 =   0   280 - 280 = 0

Third property. If you permute two parallel lines of a Matrix, its determinant changes sign.

$$\begin{pmatrix} 5 & -2 \\ 4 & 3 \end{pmatrix} \qquad \begin{pmatrix} -2 & 5 \\ 3 & 4 \end{pmatrix}$$

= 15 + 8 = 23   = -8 - 15 = -23

$$\begin{pmatrix} -2 & 4 & 5 \\ 6 & 7 & -3 \\ 3 & 0 & 2 \end{pmatrix} = -217 \qquad \begin{pmatrix} -2 & 4 & 5 \\ 3 & 0 & 2 \\ 6 & 7 & -3 \end{pmatrix} = 217$$

Fourth property. If we multiply ALL the elements of a line (row or column) of a determinant by a number other than zero (0), the determinant is multiplied by that number.

$$(3)x— \begin{pmatrix} -3 & -2 \\ 4 & 5 \end{pmatrix} \longrightarrow \begin{pmatrix} -9 & -6 \\ 4 & 5 \end{pmatrix}$$

$$= 15 + 8 = -7 \qquad\qquad = -45 + 24 = -21$$

Regularly this property is used in the opposite way, instead of multiplying, we try to factorize.

Fifth property. If a line (row or column) is added to or subtracted from another line multiplied by a number, the determinant does NOT change.

$$\begin{pmatrix} -2 & 4 & 5 \\ 6 & 7 & -3 \\ 3 & 0 & 2 \end{pmatrix} \;F_2-2F_3\; \begin{pmatrix} -2 & 4 & 5 \\ 0 & 7 & -7 \\ 3 & 0 & 2 \end{pmatrix} = 217$$

Sixth property. The determinant of a Matrix is equal to that of its Transpose.

$$\boxed{|A| = |A^t|}$$

$$A = \begin{pmatrix} 6 & 3 \\ -2 & 1 \end{pmatrix} = A^t = \begin{pmatrix} 6 & -2 \\ 3 & 1 \end{pmatrix}$$

Seventh property. If **A** has Inverse Matrix A^{-1} , then:

$$\boxed{|A^{-1}| = \frac{1}{|A|}}$$

DETERMINANT OPERATIONS

To perform operations to obtain the value of a determinant of a matrix, the matrix must always be a *Square Matrix* (2x2, 3x3, 4x4, ... etc.), i.e. *same rows same columns*.

Determinants of 2x2

Examples

$$\boxed{\det(A) = |A|}$$

$$A = \begin{pmatrix} 5 & -3 \\ 6 & 4 \end{pmatrix}$$

$|A| = 20 - (-18)$
$= 20 + 18$
$= \mathbf{38}$

$$B = \begin{pmatrix} -8 & 0 \\ 4 & -2 \end{pmatrix}$$

$|B| = 16 - (0)$
$= \mathbf{16}$

Exercises.

$$A = \begin{pmatrix} 7 & -1 \\ 8 & 12 \end{pmatrix} \qquad |A| = 84 - (-8) = \mathbf{92}$$

$$B = \begin{pmatrix} -4 & -2 \\ -7 & 5 \end{pmatrix} \qquad |B| = -20 - 14 = \mathbf{-34}$$

$$C = \begin{pmatrix} 1/2 & -2 \\ 3/4 & 4 \end{pmatrix} \qquad |C| = 2 - (-3/2) = 2/1 + 3/2 = \mathbf{7/2}$$

3X3 DETERMINANTS APPLYING THE SARRUS RULE

The **Sarrus Rule** can be used in two ways.

The first consists of writing below the matrix only the first two rows as follows:

Example 1.

$$|A| = \begin{pmatrix} -2 & 4 & 5 \\ 6 & 7 & -3 \\ 3 & 0 & 2 \end{pmatrix}$$

The *first two rows* of the same matrix are *added at the bottom*.

$$|A| = \begin{pmatrix} -2 & 4 & 5 \\ 6 & 7 & -3 \\ 3 & 0 & 2 \\ -2 & 4 & 5 \\ 6 & 7 & -3 \end{pmatrix}$$

The dominant and lower diagonals are defined.

$$|A| = \begin{pmatrix} -2 & 4 & 5 \\ 6 & 7 & -3 \\ 3 & 0 & 2 \\ -2 & 4 & 5 \\ 6 & 7 & -3 \end{pmatrix}$$

And we proceed to obtain the determinant with the following operations.

$$|A| = -28 + 0 - 36 - (105 + 0 + 48)$$
$$|A| = -64 - (153)$$
$$|A| = -64 - 153 = \textbf{-217}$$

The **Sarrus Rule** can be applied in the same way if *the first two columns are added* to the *right side* of the matrix.

$$|A| = \begin{pmatrix} -2 & 4 & 5 \\ 6 & 7 & -3 \\ 3 & 0 & 2 \end{pmatrix}$$

$$|A| = \begin{pmatrix} -2 & 4 & 5 \\ 6 & 7 & -3 \\ 3 & 0 & 2 \end{pmatrix} \begin{matrix} -2 & 4 \\ 6 & 7 \\ 3 & 0 \end{matrix}$$

$$|A| = \begin{pmatrix} -2 & 4 & 5 \\ 6 & 7 & -3 \\ 3 & 0 & 2 \end{pmatrix} \begin{matrix} -2 & 4 \\ 6 & 7 \\ 3 & 0 \end{matrix}$$

$|A| = -28 - 36 + 0 - (105 + 0 + 48)$

$|A| = -64 - (153)$

$|A| = -64 - 153 = \mathbf{-217}$

Example 2.

For the first two rows

$$|\mathbf{B}| = \begin{pmatrix} 5 & 2 & -3 \\ 0 & 8 & -1 \\ -4 & 5 & 2 \end{pmatrix}$$
$$\begin{matrix} 5 & 2 & -3 \\ 0 & 8 & -1 \end{matrix}$$

$|\mathbf{B}| = 80 + 0 + 8 - (96 - 25 + 0)$

$|\mathbf{B}| = 88 - (71)$

$|\mathbf{B}| = 88 - 71 = \mathbf{17}$

By first two columns

$$|\mathbf{B}| = \begin{pmatrix} 5 & 2 & -3 \\ 0 & 8 & -1 \\ -4 & 5 & 2 \end{pmatrix} \begin{matrix} 5 & 2 \\ 0 & 8 \\ -4 & 5 \end{matrix}$$

$|\mathbf{B}| = 80 + 8 + 0 - (96 - 25 + 0)$

$|\mathbf{B}| = 88 - (71)$

$|\mathbf{B}| = 88 - 71 = \mathbf{17}$

ANNEX

I. Define what type of matrix it is and of what dimension or order each one is.

1.

$$A = \begin{pmatrix} 0 & 0 & 0 & 0 & 0 \\ 0 & 0 & 0 & 0 & 0 \end{pmatrix}$$

2.

$$B = \begin{pmatrix} 1 & 0 & -4 & 9 \end{pmatrix}$$

*Tipo*____________ *Dimensión*________ *Tipo*____________ *Dimensión*________

3.

$$C = \begin{pmatrix} 1 \\ 0 \\ -\sqrt{8} \end{pmatrix}$$

4.

$$D = \begin{pmatrix} 1 & 2 & 3 \\ 6 & 5 & 4 \\ -3 & -4 & 0 \end{pmatrix}$$

*Tipo*____________ *Dimensión*________ *Tipo*____________ *Dimensión*________

5.

$$E = \begin{pmatrix} 1 & 0 & 0 & 0 \\ 0 & -4 & 0 & 0 \\ 3 & 4 & 5 & 0 \\ 1 & 3 & 16 & -78 \end{pmatrix}$$

6.

$$F = \begin{pmatrix} 1 & 4 & \frac{1}{3} \\ 0 & 9 & -5 \\ 0 & 0 & \pi \end{pmatrix}$$

*Tipo*____________ *Dimensión*________ *Tipo*____________ *Dimensión*________

7.

$$G = \begin{pmatrix} 1 & 0 & 0 & 0 \\ 0 & -45 & 0 & 0 \\ 0 & 0 & 3 & 0 \\ 0 & 0 & 0 & 0 \end{pmatrix}$$

8.

$$H = \begin{pmatrix} 1 & 0 & 0 \\ 0 & 1 & 0 \\ 0 & 0 & 1 \end{pmatrix}$$

*Tipo*____________ *Dimensión*________ *Tipo*____________ *Dimensión*________

II. solve the following matrix operations exercises.

1.

$$\begin{pmatrix} 0 & -1 \\ -4 & -2 \\ 3 & -9 \end{pmatrix} + \begin{pmatrix} 0 & 1 \\ 4 & 2 \\ -3 & 9 \end{pmatrix}$$

2.

$$\begin{pmatrix} 2 & 1 & 3 \\ -4 & 2 & 1 \end{pmatrix} - \begin{pmatrix} 2 & 0 & 4 \\ 3 & 2 & 5 \end{pmatrix}$$

3.

$$\begin{pmatrix} 3-a & b & -2 \\ 4 & -c+1 & 6 \end{pmatrix} + \begin{pmatrix} 2 & a+b & 4 \\ 1-c & 2 & 0 \end{pmatrix}$$

4.

$$\begin{pmatrix} x-y & -1 & 2 \\ 1 & y & -x \\ 0 & z & 2 \end{pmatrix} + \begin{pmatrix} y & 0 & z \\ -z & 2 & 3 \\ -2 & 3 & x \end{pmatrix}$$

5.

$$-5 \cdot \begin{pmatrix} 2 & 1 & 3 \\ -4 & 2 & 1 \end{pmatrix}$$

6.

$$\begin{pmatrix} -3 & 2 & 1 & 4 \\ 2 & 5 & 3 & -2 \end{pmatrix} \cdot \begin{pmatrix} 0 & -4 & 1 \\ 1 & -2 & 1 \\ 2 & 0 & 2 \\ 3 & 2 & 1 \end{pmatrix}$$

7. International support, in millions of dollars, from three leading economic countries *A*, **B** and *C* to three other developing countries **X**, **Y** and **Z**, during the pandemic years 2019 and 2020 is given by the information in the following matrices:

$$A_{2019} = \begin{matrix} & X & Y & Z \\ A \\ B \\ C \end{matrix} \begin{pmatrix} 11 & 6'7 & 0'5 \\ 14'5 & 10 & 1'2 \\ 20'9 & 3'2 & 2'3 \end{pmatrix} \qquad A_{2020} = \begin{matrix} & X & Y & Z \\ A \\ B \\ C \end{matrix} \begin{pmatrix} 13'3 & 7 & 1 \\ 15'7 & 11'1 & 3'2 \\ 21 & 0'2 & 4'3 \end{pmatrix}$$

Calculate and express in matrix form the total support received by each country.

How many millions did country Z receive from country B?

Calculate the total amount of support from the leading countries to each country.

8. For the following matrices A and B, calculate B-A.

$$A = \begin{pmatrix} 1 & -3 \\ -2 & 6 \end{pmatrix}, \quad B = \begin{pmatrix} 3 & -5 \\ 2 & 1 \end{pmatrix}$$

If the product A-B is performed, would the resulting matrix be the same?

9. As in the previous exercise, determine whether B-A is the same as A-B.

$$A = \begin{pmatrix} 1 & -1 \\ 0 & -2 \\ 4 & 1 \end{pmatrix}, \quad B = \begin{pmatrix} 3 & 0 & 2 \\ 1 & -1 & 5 \end{pmatrix}$$

10. Calculate all possible products of the following 3 matrices.

$$A = \begin{pmatrix} 1 & 2 & 3 \\ 1 & 1 & 1 \\ 0 & 2 & -1 \end{pmatrix} \quad B = \begin{pmatrix} 1 \\ 2 \\ 1 \end{pmatrix} \quad C = \begin{pmatrix} 2 & 1 & 0 \\ 3 & 4 & 5 \end{pmatrix}$$

11. Determine the transposes of the following matrices.

$$A = \begin{pmatrix} 2 & 1 & 0 & 7 \\ -3 & 4 & 2 & 1 \end{pmatrix} \quad B = \begin{pmatrix} 1 & 1 & 2 \\ 2 & 0 & -1 \\ -6 & -1 & 0 \end{pmatrix} \quad C = \begin{pmatrix} 1 & 3 & 3 \\ 1 & 4 & 3 \\ 1 & 3 & 4 \end{pmatrix}$$

12. For the matrices

$$A = \begin{pmatrix} 1 & -1 & 2 \\ 4 & 0 & -3 \end{pmatrix} \quad B = \begin{pmatrix} 0 & 3 & 4 \\ -1 & -2 & 3 \end{pmatrix} \quad C = \begin{pmatrix} 2 & 3 & 0 & 1 \\ -5 & 1 & 4 & -2 \\ 1 & 0 & 0 & -3 \end{pmatrix} \quad D = \begin{pmatrix} 2 \\ 1 \\ 3 \end{pmatrix}$$

calculate:

a) $A + B$ **b)** 3A-4B   **c)** A -B   **d)** A -D   **e)** B -Cf) C -D

g) A^t -C   **h)** D^t -At   **i)** B^t -A   **j)** D^t -D   **k)** D -D^t **l)** 3C-C

13. Obtain the inverse matrix of the following matrices.

a.

$$A = \begin{pmatrix} 1 & 2 \\ -1 & 1 \end{pmatrix}$$

b.

$$B = \begin{pmatrix} 1 & 1 & 0 \\ -1 & 1 & 2 \\ 1 & 0 & 1 \end{pmatrix}$$

c.

$$C = \begin{pmatrix} 1 & 2 & -3 \\ 3 & 2 & -4 \\ 2 & -1 & 0 \end{pmatrix}$$

d.

$$D = \begin{pmatrix} -2 & 1 & 4 \\ 0 & 1 & 2 \\ 1 & 0 & -1 \end{pmatrix}$$

14. From matrix **A** obtain the complementary minors.

$$A = \begin{pmatrix} -2 & 4 & 5 \\ 6 & 7 & -3 \\ 3 & 0 & 2 \end{pmatrix}$$

And then obtain the adjoining matrix of **A**

15. Let **A** and **B** be the following square matrices of dimension 2×2:

$$A = \begin{pmatrix} 3 & -1 \\ 1 & 0 \end{pmatrix} \qquad B = \begin{pmatrix} 4 & 2 \\ -1 & 3 \end{pmatrix}$$

Calculate the matrix X that verifies the following matrix equation:

$$AX = B$$

16. Let **A**, **B** and **C** be the following matrices of order 2:

$$A = \begin{pmatrix} 3 & 6 \\ 2 & -1 \end{pmatrix} \qquad B = \begin{pmatrix} -2 & 1 \\ 3 & -3 \end{pmatrix} \qquad C = \begin{pmatrix} 6 & 4 \\ 3 & -2 \end{pmatrix}$$

Calculate the matrix X that verifies the following matrix equation:

$$A + XB = C$$

17. Let **A**, **B** and **C** be the following matrices of order 2:

$$A = \begin{pmatrix} -1 & 1 \\ 1 & 0 \end{pmatrix} \qquad B = \begin{pmatrix} 4 & -2 \\ 1 & 0 \end{pmatrix} \qquad C = \begin{pmatrix} 6 & 4 \\ 22 & 14 \end{pmatrix}$$

Calculate the matrix X that verifies the following matrix equation:

$$AXB = C$$

18. Let **A** and **B** be the following matrices of dimension 3x3:

$$A = \begin{pmatrix} 1 & 0 & 1 \\ 0 & -1 & 0 \\ 1 & 2 & 2 \end{pmatrix} \qquad B = \begin{pmatrix} 1 & -1 & 0 \\ 2 & 3 & -2 \\ -3 & 1 & -1 \end{pmatrix}$$

Calculate the matrix X that verifies the following matrix equation:

$$B^t - AX = B$$

19. Calculate the following determinants.

a)

$$\begin{vmatrix} 1 & 3 \\ -1 & 4 \end{vmatrix}$$

b)

$$\begin{vmatrix} -2 & -3 \\ 2 & 5 \end{vmatrix}$$

20. From the following matrices calculate the corresponding determinants.

$$\begin{pmatrix} 1 & 8 & 1 \\ 1 & 7 & 0 \\ 1 & 6 & -1 \end{pmatrix} \qquad \begin{pmatrix} 3 & 4 & -6 \\ 2 & -1 & 1 \\ 5 & 3 & -5 \end{pmatrix} \qquad \begin{pmatrix} 7 & 8 & 0 \\ 0 & -7 & 3 \\ 1 & 0 & 1 \end{pmatrix} \qquad \begin{pmatrix} 0 & 3 & 1 \\ -2 & 0 & 2 \\ 3 & 4 & 0 \end{pmatrix}$$

BIBLIOGRAPHY

Baldor, A. (2019). *Álgebra.* Mexico: Patria.

CURSIN. (July 25, 2023). *CURSIN.* Retrieved from https://cursin.net/curso-de-algebra-lineal-para-principiantes-de-todas-las-edades/

Estrada Coronado, R. M. (2019). *Álgebra.* Mexico: Pearson.

FACIALIX (July 25, 2023). *FACIALIX.* Retrieved from https://blog.facialix.com/curso-gratis-en-espanol-de-algebra-lineal/

Grossman (2019). *Algebra Lineal.* Mexico: Mc Graw Hill.

Guzmán, F. (2011). *Álgebra lineal: Serie universitaria* (1st ed.). Mexico: Patria.

Hernández Pérez, M. (2021). *Álgebra Lineal. Ejercicios de Práctica (*2nd ed.). México: Pearson.

Larson, R. (2013). *Fundamentals of Linear Algebra* (7th ed.). Mexico: Cengage Learning.

Lay, D. C. (2007). *Linear Algebra and its applications.* Mexico: Pearson Educación.

Salazar Guerrero, L. J., & Bahena Román, H. (2021). *Álgebra* (1st ed.). México: Patria Educación.

Wikipedia (July 28, 2023). *Wikipedia.* Retrieved from https://es.wikipedia.org/wiki/%C3%81lgebra_lineal